AF497271

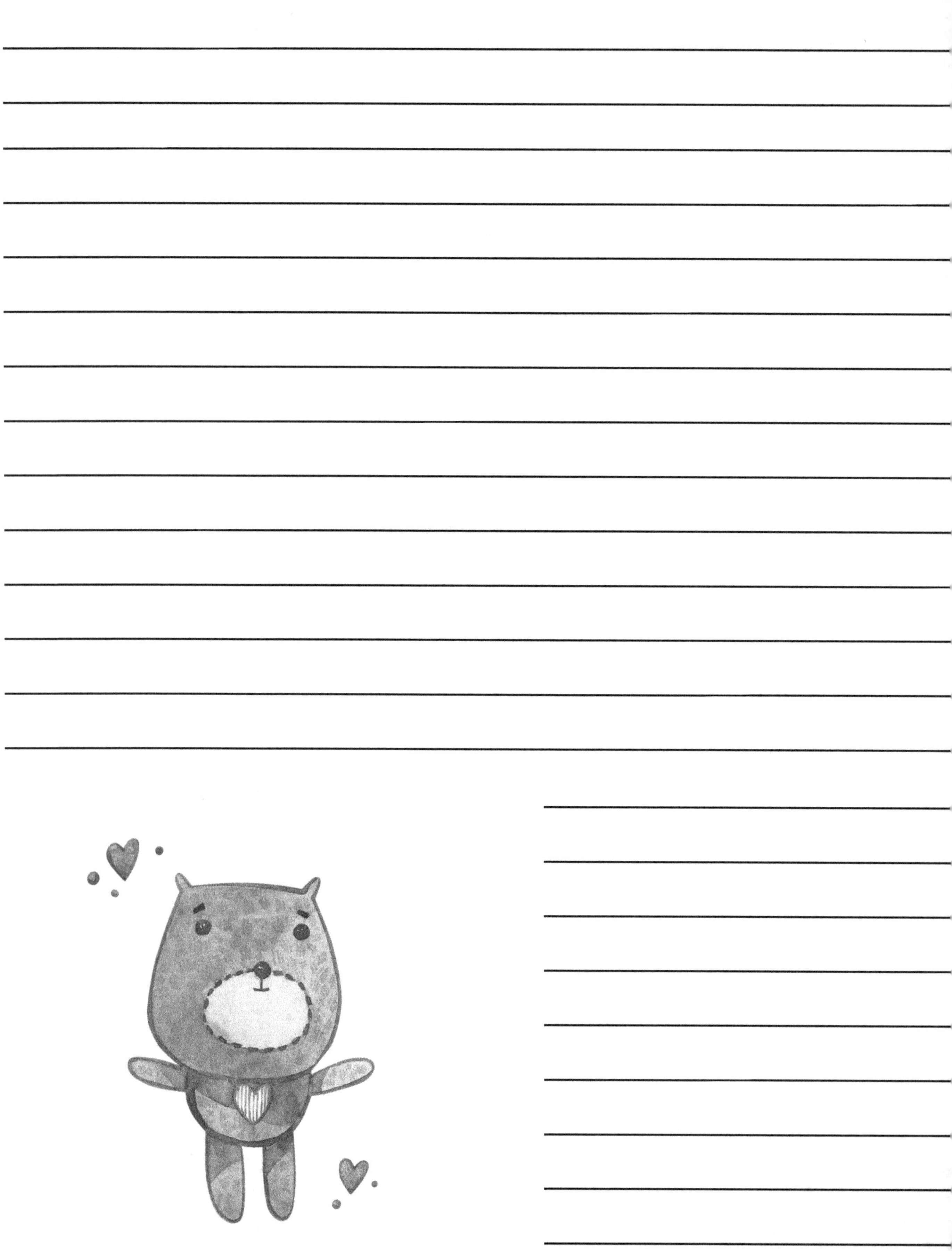

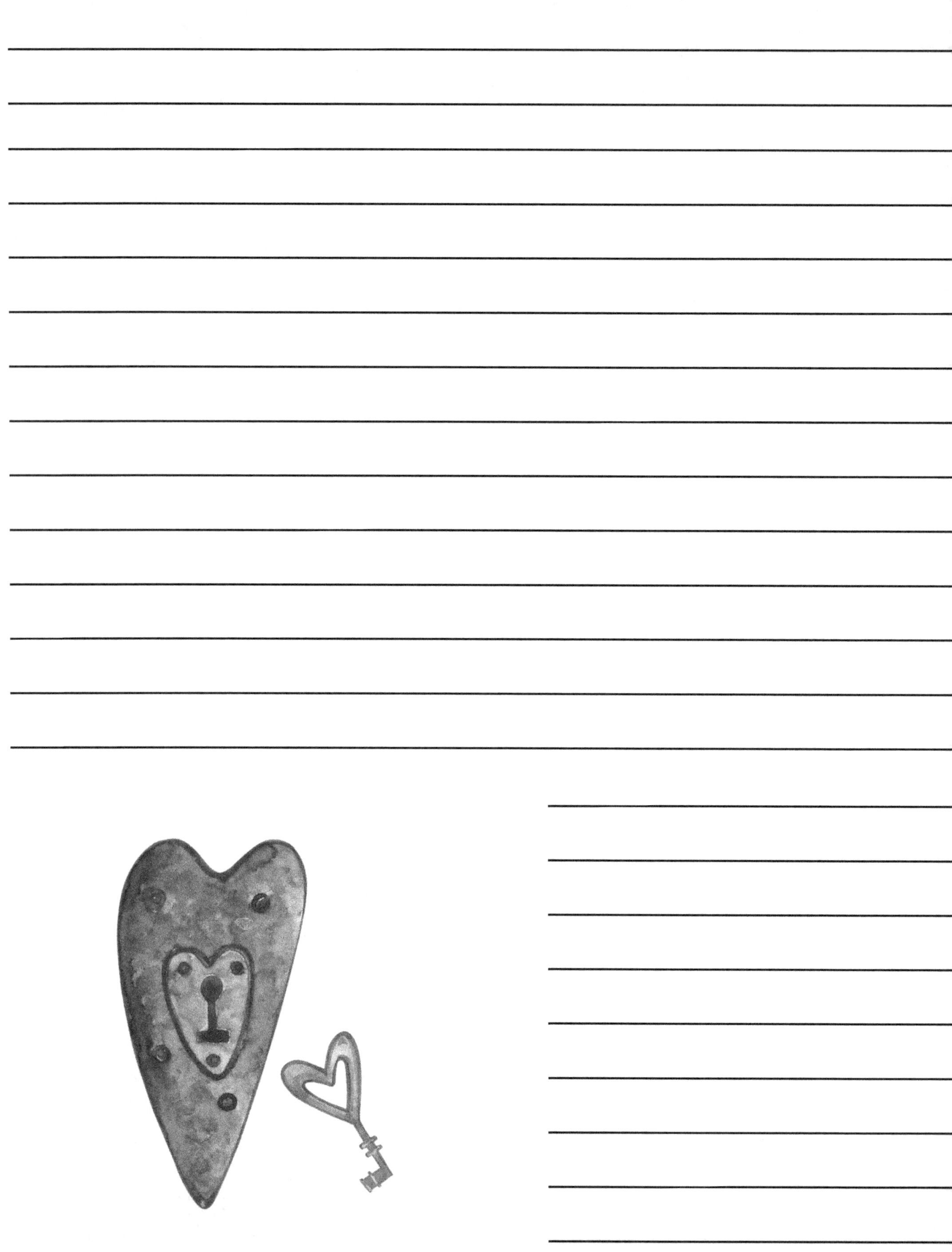

you
+
me
=
love

i love
you

you are loved

love
is all
you
need

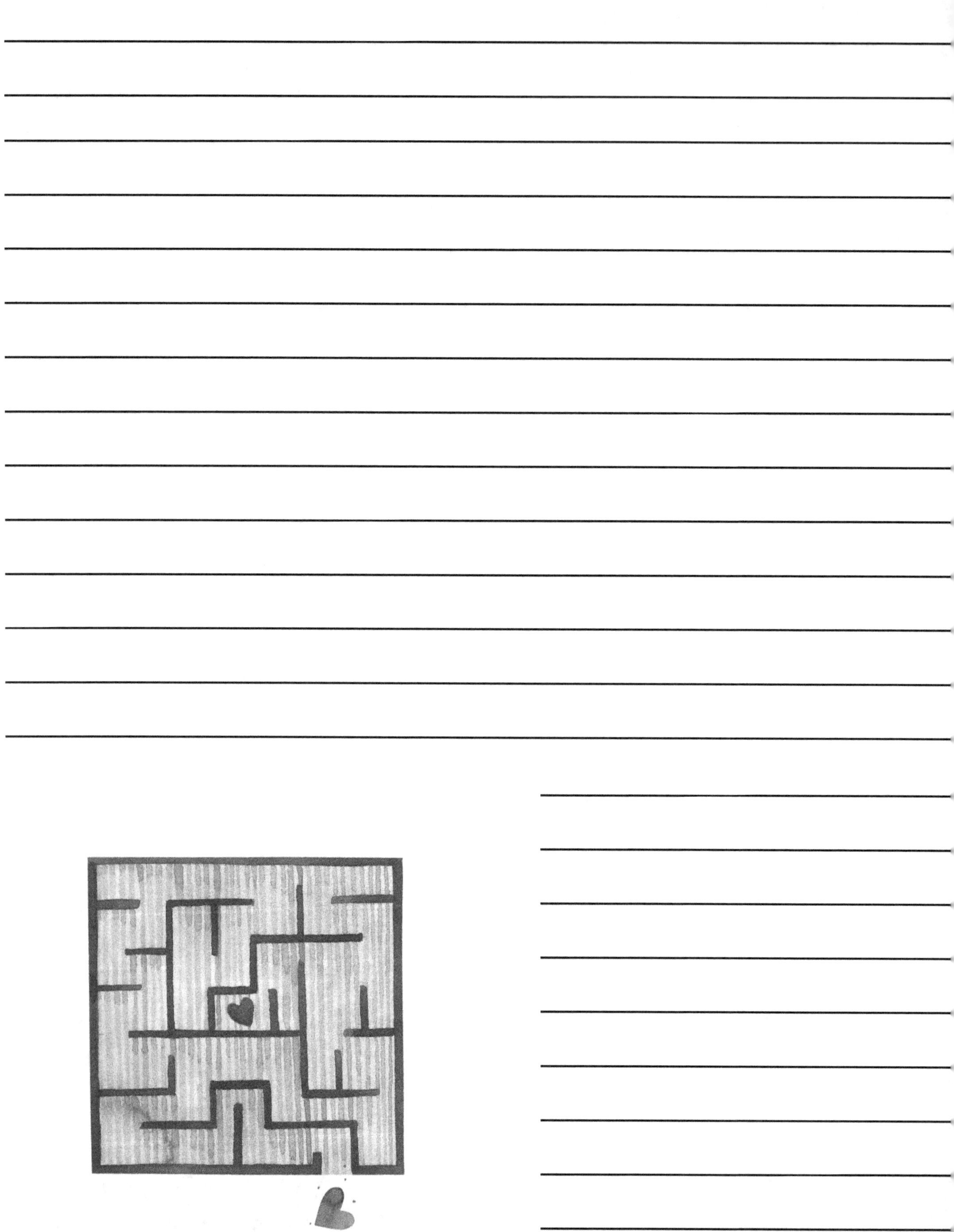

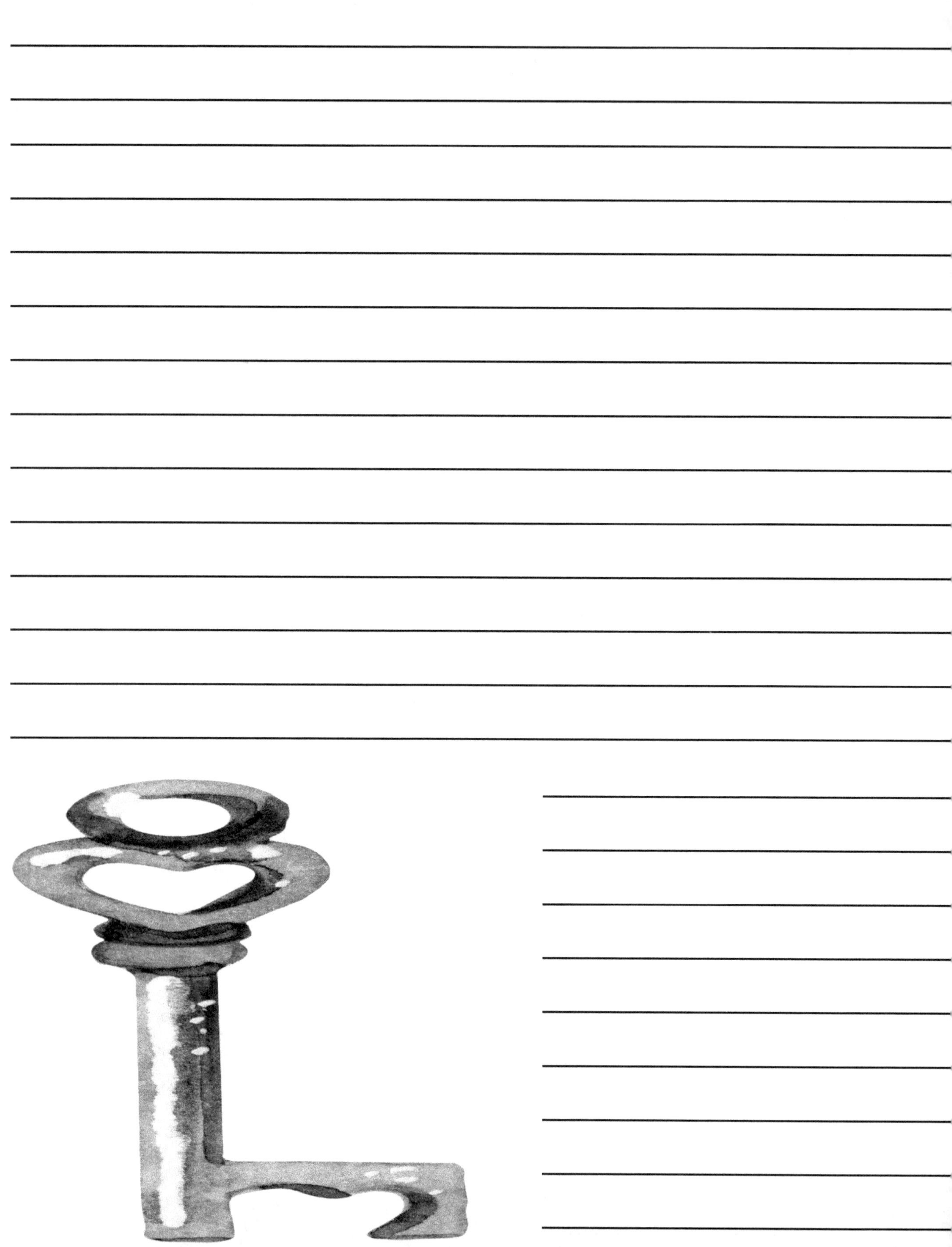

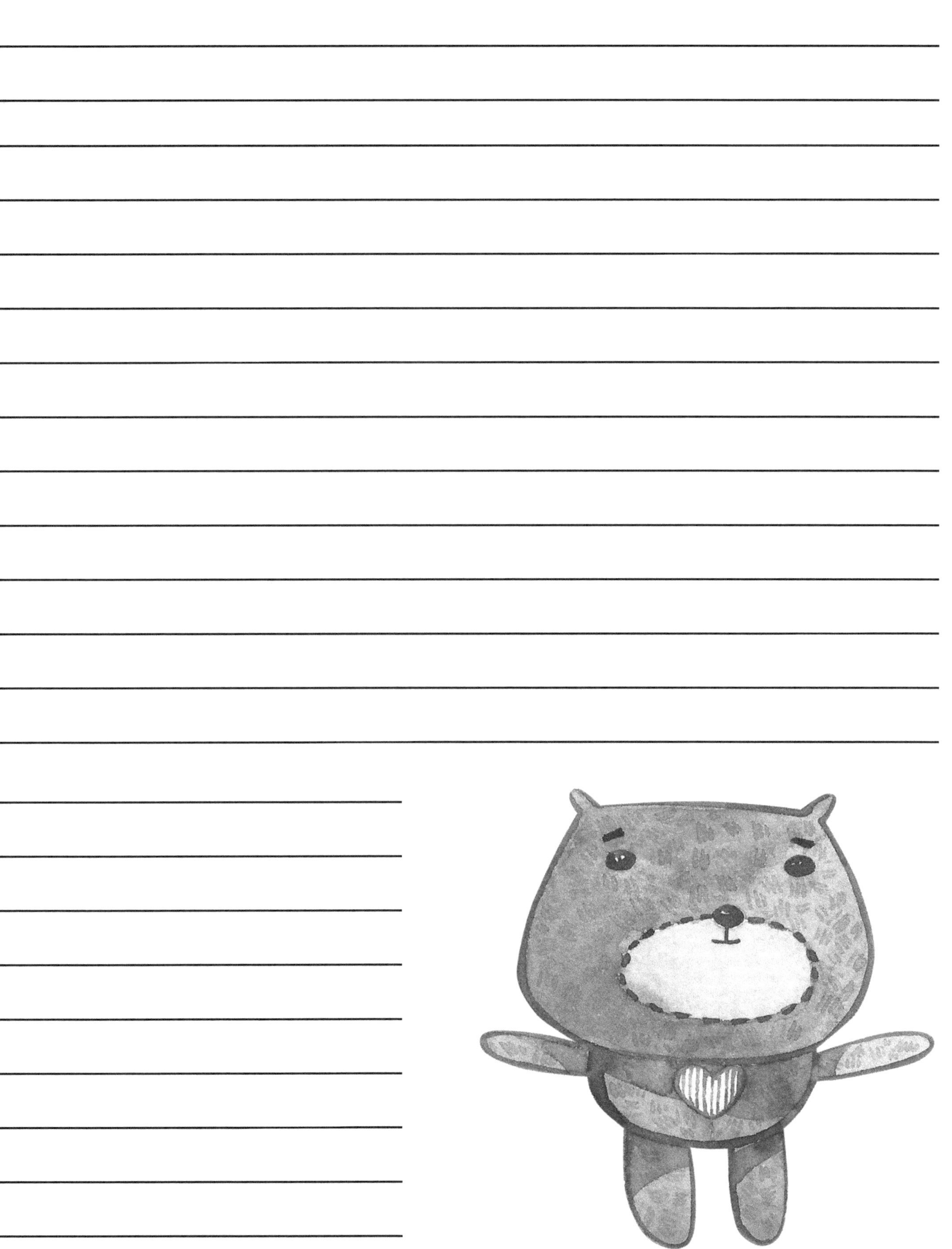

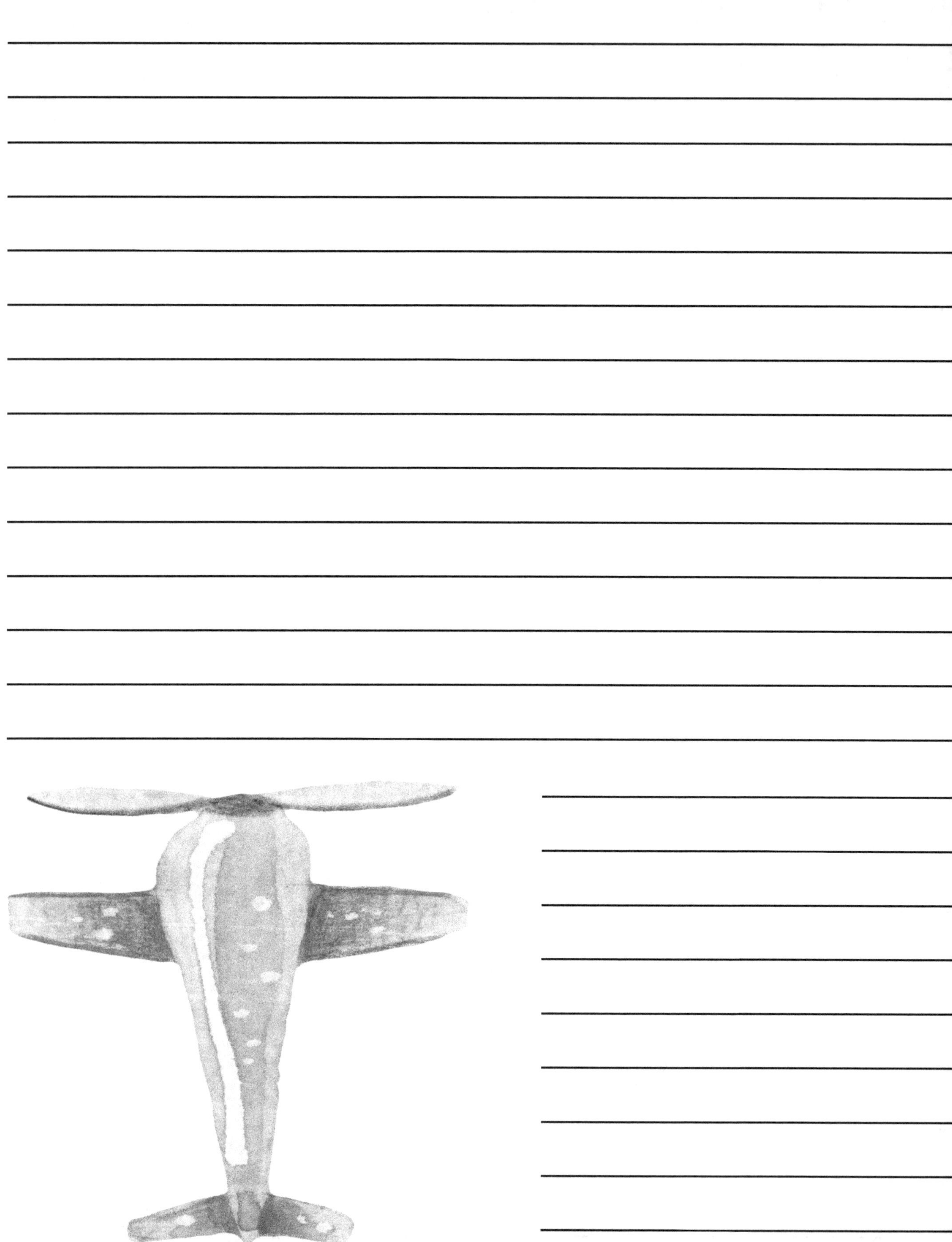

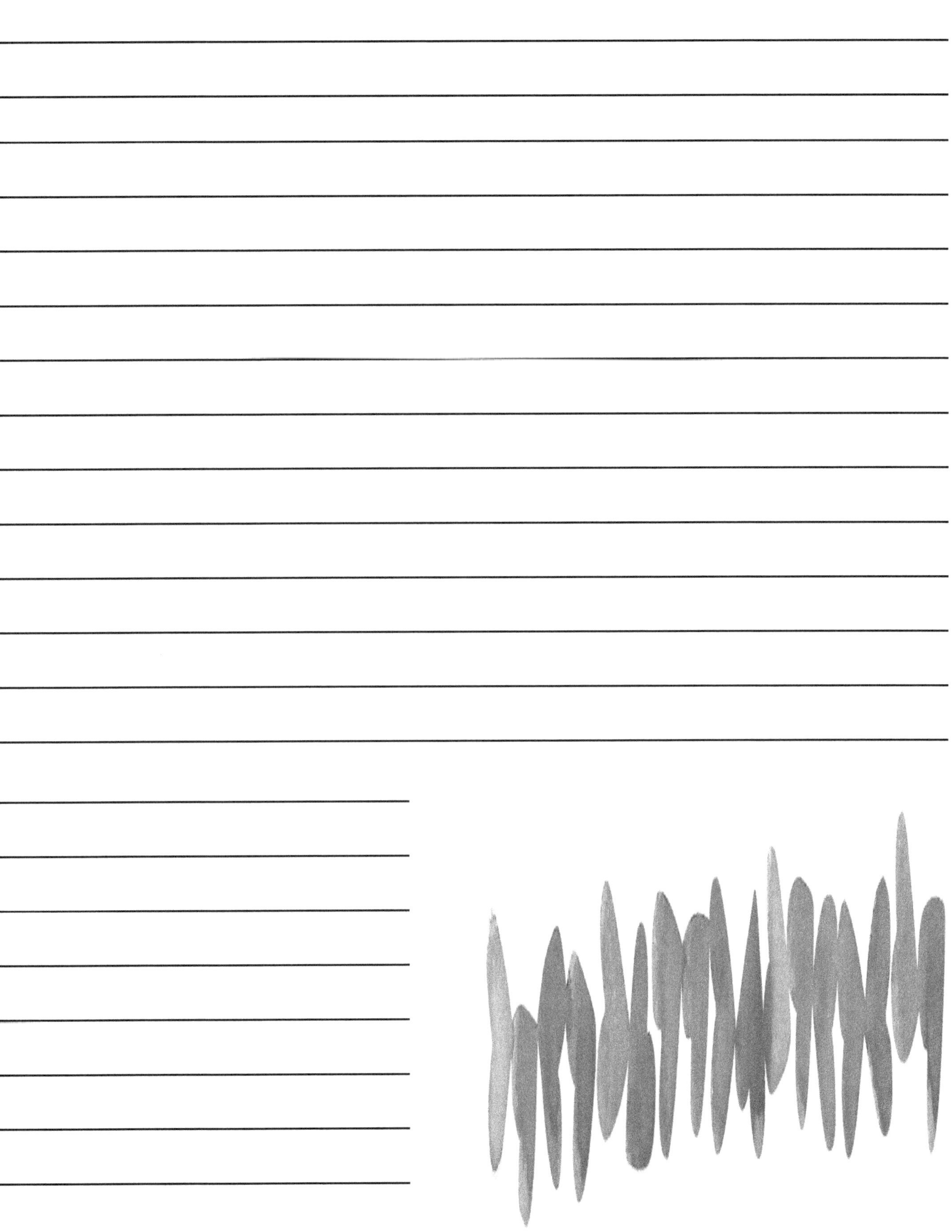

you + me = love

believe
IN
Love

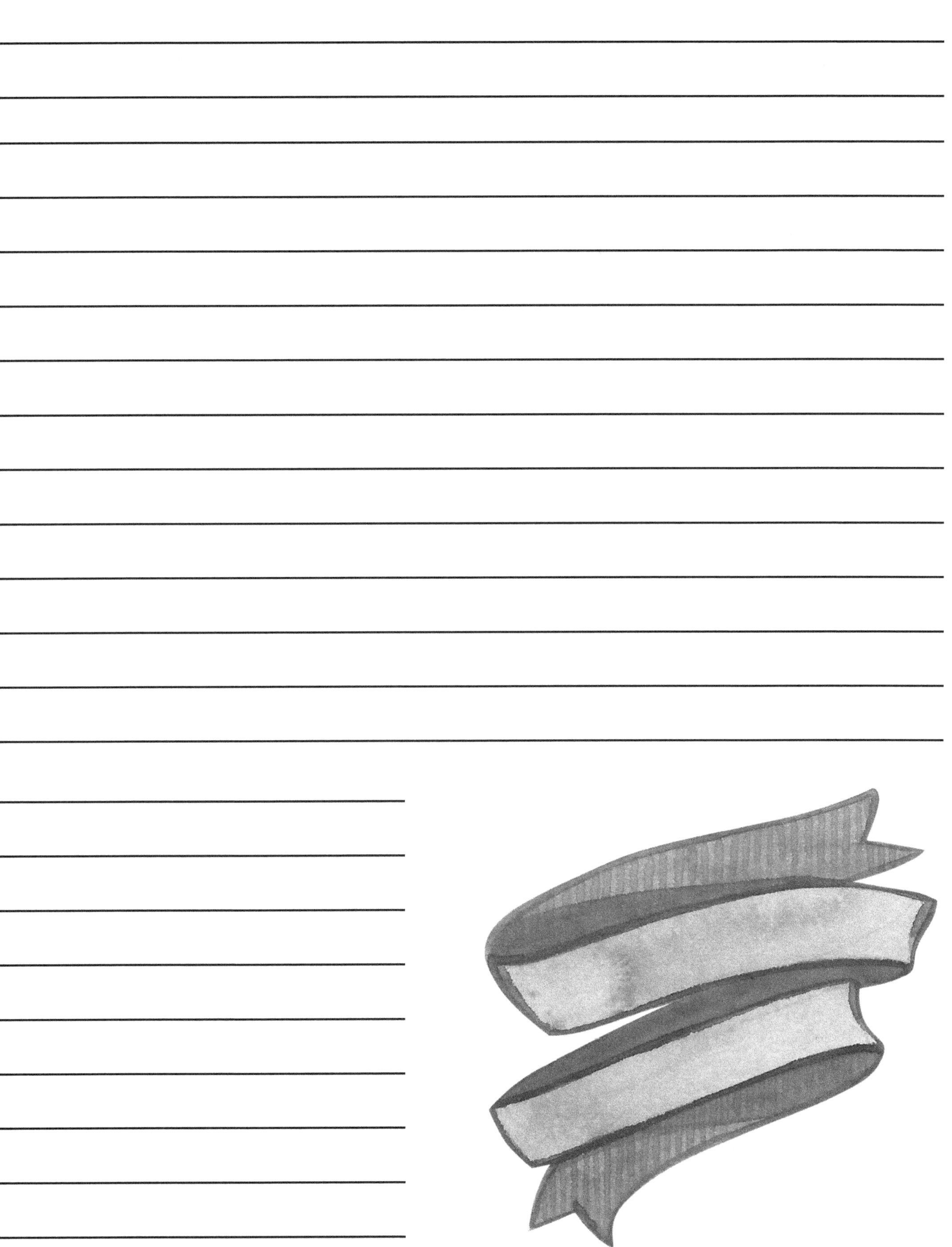

i love
you

you are loved

love
is all
you
need

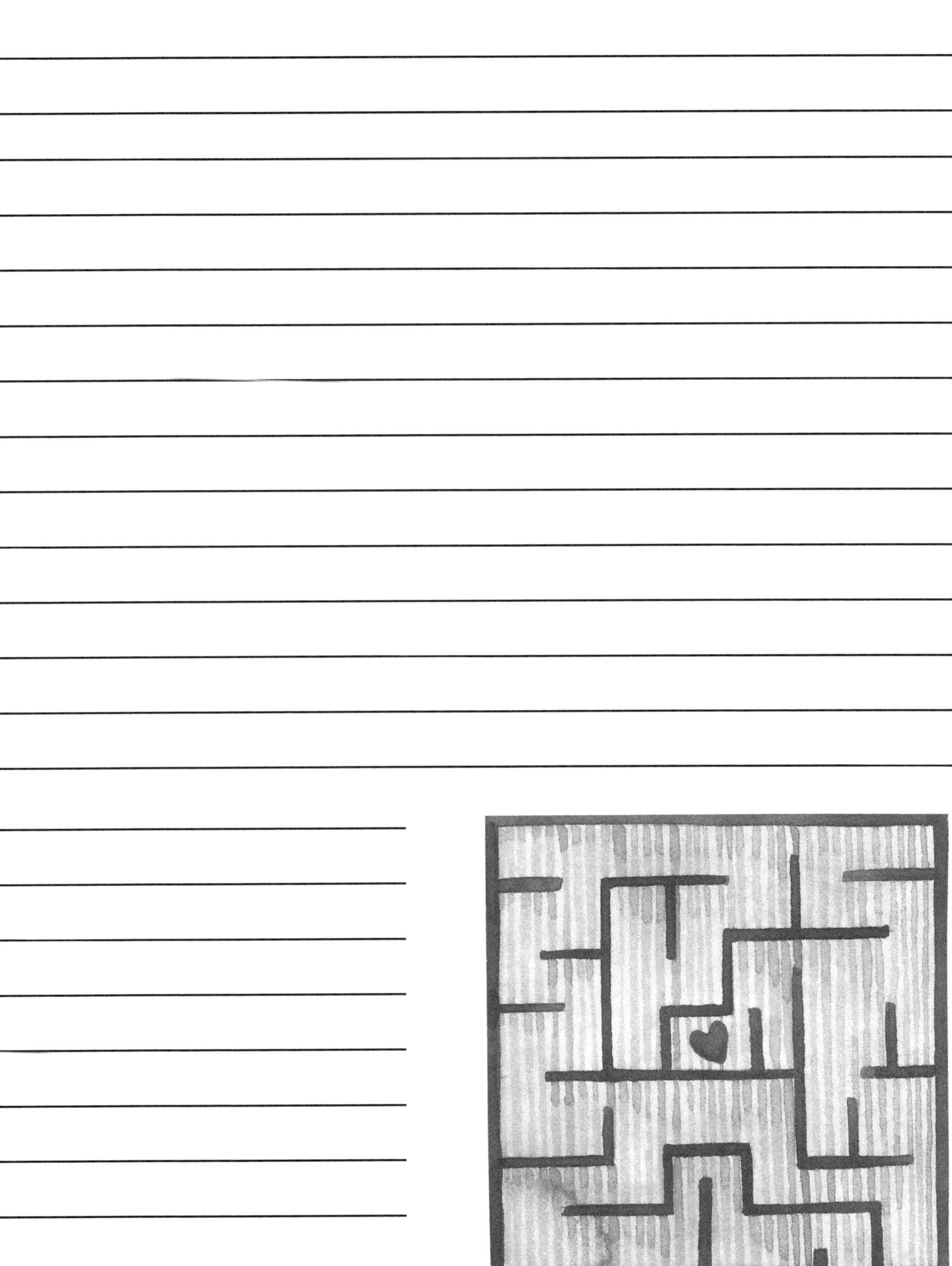

- Acts of Kindness Tracker -

- Acts of Kindness Tracker -

- Acts of Kindness Tracker -

Notes

Notes

Notes

www.ingramcontent.com/pod-product-compliance
Lightning Source LLC
LaVergne TN
LVHW080558200726
843510LV00004B/947